YOUR KNOWLEDGE HAS VALUE

- We will publish your bachelor's and
  master's thesis, essays and papers

- Your own eBook and book -
  sold worldwide in all relevant shops

- Earn money with each sale

Upload your text at www.GRIN.com
and publish for free

Bibliographic information published by the German National Library:

The German National Library lists this publication in the National Bibliography;
detailed bibliographic data are available on the Internet at http://dnb.dnb.de .

Imprint:

Copyright © 2016 GRIN Verlag, Open Publishing GmbH
Print and binding: Books on Demand GmbH, Norderstedt Germany
ISBN: 9783668435087

This book at GRIN:

http://www.grin.com/en/e-book/358117/a-theory-of-the-universe-and-matter

Daniel Stark

A Theory of the Universe and Matter

GRIN Publishing

A Theory of the Universe and Matter

Daniel Lee Stark
April 7, 2016

Abstract

This paper presents a new concept for particle physics, which leads to a new concept for the universe and theories in physics. That concept is that the fundamental particle in the universe is a Dark Matter (DM) particle, and our matter consists of excited DM particles. The excited DM particles are theorized as formed in small regions of the universe where DM particles have been concentrated via gravitational attraction, which also acts to concentrate gravitational potential energy. These high-energy DM concentrations energize DM particles, creating the electromagnetic fields, which forms our matter and our sub-universe. The DM particles only interact via a gravitational field; however, the excited DM is theorized to also create the electromagnetic force. The weak and strong nuclear forces are theorized as an edifice of the DM basic bubble like structure. All matter is composed of only two items, DM and energy. All particles are composed of different combinations of these two items.

The greater universe is an even distribution of DM particles with all energy stored as gravitational potential. Our sub-universe is a concentration of DM and energy. The energy flows back to lower energy regions in the universe following the same principle as the thermodynamic laws. The energy flows back to the universe's lower energy regions via excited DM particles, creating the appearance of an expanding universe.

Thus we live in two universes, an infinite DM universe and a finite sub-universe composed of our matter with both universes sharing the same space, time, energy, gravity and DM particles.

This theory has major implications in "a theory of the universe," particle physics and general physics. This simple assumption leads to contradictions to many of current accepted theories. The model presented herein is less complex than the current Standard Model, origin of the universe, and relativity. Many current contradictions are resolved. This theory enhances mathematical modeling of physical phenomena, enabling a more rigorous mathematical treatment for physics in general.

A Theory of the Universe and Matter

Table Of Contents

A Theory of the Universe and Matter

1. Logic Used to Derive This Theory

1.1. Recent Revelations About Our Universe and Dark Matter

Prior to providing the theory advanced by this paper, the evidence and the logic that derived this theory is presented.

First, the evidence collected in recent astronomical discoveries forms the framework for this theory. Abbreviated samples of this evidence is referenced and partially presented below.

(a) An image of galaxy MCS J0416.1-2403, one of six clusters targeted by the Hubble Frontier Fields program, is shown in figure one. [1] The image's blue is a DM map detected by Hubble combined with Chandra using gravitational lensing.

Figure 1 DM Map of MCS J0416.1-2403

DM appears more evenly spaced in the universe, and around new galaxies whereas our matter is mainly in galaxies, with little matter between galaxies. Our Universe is distributed in a highly structured fashion. Large concentrations of stars and cosmic dust interspersed with large areas of empty space. These star concentrations form galaxies, with very little of our matter between galaxies, empty space. The bigger picture shows galaxies in clusters. Galaxy clusters are located in the densest concentrations formed in large strings. This cosmic network of strings shows very little of our matter between them. Matter appears to first concentrate into large strings, then into galaxies.

(b) The Super Massive Black Hole (SMBH) at the center of galaxy NGC 1365 has a disk edge velocity at 84% the speed of light. [2] Such a SMBH, if typical, creates a very high-energy particle accelerator. When the edge of the SMBH is spinning at 84% of the speed of light, particle escape is easily achieved. The escape velocity adds to the Eddington luminosity.

The faster the black hole spins, the larger its ergo-sphere, which provides high-energy collisions farther from the event horizon adding to the probability that a particle will escape into the galaxy.

(c) Galaxies are observed to have SMBHs at the center with new star formation near the center of these galaxies. Galaxies without these SMBHs do not have new star formations.

Bogdan and his colleague Andy Goulding from Princeton University investigated the correlation between SMBHs and DM halos around galaxies. [3] They used data from 3,000 elliptical galaxies. Star motions provided the mass for the galaxy's central black hole, and X-ray measurements of hot gas surrounding the galaxies determined the DM halo's mass.

DM constitutes most of the matter in our observable universe, possibly as much as 90%. Every galaxy is surrounded by a halo of DM that weighs as much as a trillion suns and extends for hundreds of thousands of light-years. A strong correlation was determined between the DM halo's mass and the black hole's mass. [4]

(d) Researchers established that a galaxy's evolution and structure is directly related to the size of the massive black hole(s) in the center of the galaxy. A second observation is that star revolution speed at the outer edge of the galaxy was directly related to the size of the galaxy's inner black hole.

(e) New suns and solar systems are born near the center of each galaxy, and propagate outward over time. Matter rotating in the same direction as the galaxy's SMBH allows formation of structured solar systems with sufficient angular momentum to prevent the matter from returning to the SMBH.

(f) The most spectacular accretion disks found in nature are those of active galactic nuclei and of quasars, which are believed to be Massive Black Holes (MBH) at the center of galaxies. The accretion disk is the matter above the event horizon around the MBH.

1.2. A New Theory for the Nature of Particles Based on Logic

The second part of the logic leading to the derivation of this theory requires speculation about the nature of matter, and identification of the fundamental particle. Selecting a DM particle as the basic particle is based on the realization that our matter is possibly only 10% of the universe, and may be the exception rather than the rule.

1.2.1. Our Matter Consists of DM Excited Particles

The first step is to understand the nature of our particles. Our intuitive senses provide a concept of "solid" which prevents particles from passing through each other, but bouncing upon collision. However, particles have repulsive forces from electromagnetic fields, which creates the illusion of "hard particles" upon collision.

The concept that a particle is mainly open space is not new. Literature sources cite early German scientists referring to matter as froth, with even Einstein possibly describing matter accordingly. In 1955 John Wheeler described matter as Quantum foam (also referred to as space-time foam). [5]

The theory presented herein theorizes the DM particles as a gravity bubble. The concept of a gravity bubble is selected herein rather than a graviton or a wimp in order to free the DM particle from properties currently assigned to those defined particles. However, a graviton may be more appropriate if its properties are redefined. A graviton's current properties are described as: massless, relativistic velocities, a particle of force, no weak interactions.

A bubble structure is selected to avoid the gravitational force approaching infinity as one approaches the "center" of the DM particle. The minimum dimension is most likely Plank's constant, which avoids excessive particle gravitational fields near a particle's center. The lesson is that nature does not have singularities or points. Points are an edifice of math, not represented by matter in nature.

The bubble structure is oversimplified in that a better analogy may be described as a hollow liquid droplet, which does not have a surface, but a decreasing fluidic nature radially outward. A "bubble" structure with an infinite number of shells may suffice as an adequate mathematical model. A shell is described by the allowed energy quanta. The bubble structure is similar to a balloon wherein quantized energy packets in a shell are allowed; however, the number of allowed shells is infinite. Each shell can support three modes, the stable mode being (1,1,1). This complex model will be referred to in its simpler form as a "gravity bubble" in this paper.

1.2.2. DM Particles Have Mass & Mass is Conserved

A DM particle satisfies the definition of mass because it is capable of supporting kinetic energy and gravitational attraction, thus having inertial mass and gravitational mass.

All particles will have mass since all particles are either a ground state DM or an energized DM particle. The revelation is that mass can be applied to a DM particle, and an energized DM has the same mass as the original DM. Mass and energy are conserved even during nuclear reactions. This theory contradicts the current graviton description by assigning mass to the gravity bubble, thus requiring definition of a new DM particle. Weak interactions are not ruled out for a DM particle in the "bubble" theory advanced herein. The DM particle is discussed in more depth later in this paper.

1.2.3. DM Bubble Structure Can Host Energy

This bubble theory adds to the "froth" concept by identifying a DM particle as capable of being excited, and creating our known matter. The "bubble" like structure describes a structure that can support energy, thus excite the DM particle. The DM particle is described as a bubble structure that has mass, and supports energy in several degrees-of-freedom such as three vibration modes each on possibly multiple shells, with each shell having up to two spins. The DM particle can also host kinetic energy. A very small subset or all DM excitations are expected to be stable, and possibly only a subset of excitation modes creates the electromagnetic force creating our familiar particles.

1.2.4. DM Particles May Also Combine & Group

The DM particles are also expected to combine and/or group, further increasing the potential particle types and complexity of matter. The binary nature of the universe consists of energy and DM particles, but the combinations of DM particles merging and grouping as well as energy supported on many Degrees-of-Freedom (DoF) creates a large number of possible combinations and different particles.

A current particle theory describes DM as a weakly interacting massive particle (WIMPs). When two WIMPs collide they annihilate producing an electron and an antimatter positron. Positrons and electrons are identical except for opposite electrical charges.

This theory chooses to describe the DM particle more like a bubble, thus from now on referred to as the "bubble theory." This DM bubble structure can be excited in many DoF, creating our particles. Possible supporting evidence is seen in electron microscope pictures of atoms, which reveal a strong similarity to standing wave patterns. These standing waves inside the particle's volume provide structure. Thus hard discreet particle like behavior is actually caused by repulsive and attractive forces such as the electromagnetic force.

The DM particle may be described as a 3-D string, bubble like structure rather than a two-dimensional "string," with the concept that the DM particles may also combine or group, forming composite bubble like structures of different diameters.

For the purposes of this paper a DM particle is designated as "G" with a ground state DM particle designated as G_g and an excited DM particle as G_e. The term G includes both G_g and G_e.

1.2.5. Summary of The Bubble Theory

Summarizing, logic derivation presented above leads to the theory that all matter is composed of the basic Dark Matter (DM) particles, and our matter is an excited version of DM, with the excited DM creating the Electromagnetic (EM) force. This simple concept provides a new description of our universe, and the nature of matter. This concept is less complex than the current Standard Model, which is an indication that the bubble theory is correct.

Energy is quantized on this bubble structure; however, the quantized energy may change by moving into another "shell" on the bubble structure. Thus the energy that can be accepted on a DM vibration mode can be any value within the allowed shell boundaries by selecting the allowed shell.

2. New Concept for the Universe

2.1. Our Sub-Universe Inside an Infinite DM Universe

Our universe consists of two universes within each other, a DM universe, and an excited DM (our matter) sub-universe. Both universes share the same space, time, energy, DM, and gravitational fields.

However, the DM universe extends to infinity, and is mainly an even distribution of DM particles at rest with all

energy stored as gravitational potential. Our universe is
created by a disturbance in the DM universe, which caused an
accumulation of DM material. These DM concentrations are finite
in space and time, and result in our matter being created
concentrated in a finite space and time. Our universe is
contained in the larger DM universe, and multiple universes like
ours are expected to exist in the infinite DM universe. These
sub-universes are born and die in the infinite DM universe.

2.1.1. Ground State DM Universe

The ground state DM universe's energy is stored as
gravitational potential energy. That energy is evenly
distributed in the DM universe, with the gravitational force on
each DM particle nulled by the even distribution and symmetry of
motionless DM particles. This is the lowest energy state
possible, or ground state for the DM universe, with all of the
energy stored as gravitational potential.

2.1.2. Creation of Our Sub-Universe

Fluctuations in this even DM distribution, caused by
unbalanced gravitational attractions, create concentrations of
DM particles and the gravitational potential energy. The DM and
energy concentrations create "hot spot" energy regions. These
high-energy concentrations first transfer the gravitational
potential energy into DM kinetic energy, which acts to excite DM
particles possibly via high-speed collisions, creating our
observable matter and our observable universe.

2.1.3. Expanding Universe

The high energy concentration is unstable, and will
redistribute to "lower energy" regions of the universe. The
energy redistributes itself to the lower energy state universe
by hitching a ride on our energized DM particles. Thus our
universe is finite in both space and time, contained in a DM
universe infinite in space and time. The universes are a
competition between gravity and energy, wherein fluctuations in
the DM distribution allow for gravitational driven accumulations
of DM particles, which also concentrates energy. The energy
drives the DM concentration back into the lower energy regions
of the DM universe.

The DM universe is not expanding, but our matter is
expanding back into the infinite stable DM universe, creating
the impression of an expanding universe. Our excited DM sub-

universe is expanding driven by Dark Energy (DE) which is the original gravitational potential stored between the distributed DM particles.

The energy thus hosted on excited DM particles flow the energy back into lower energy regions in the universe. This process is consistent with the concept of the first rule of thermodynamics. The first law states: "In any process, the total energy of the universe remains constant."

Our sub-universe is expanding at the speed of light back into the infinite DM universe, transported outward on the excited DM particles (photons). Space is not expanding, but our excited matter sub-universe is expanding back into the infinite DM universe. The excited DM particles lose their energy back into gravitational potential transforming back into evenly distributed DM particles. This theory explains the expanding universe and identifies the DE.

This sub-universe description is consistent with the current "Tired-light" proponents, Thomas Van Flandern, PhD, and theoretical physicist Paul LaViolette. LaViolette estimates that "a photon (excited DM's) energy loss of just 10% for every billion years of its travel would entirely account for the observed increase of red-shift with distance. The sum of actual astronomical data fits the tired-light model better than the expanding universe model. [6]

2.1.4. DM Galaxy Formation

Current galaxy formation theories describe the early universe composed of our matter and dark matter. The current theory describes our matter as collecting via gravitational attraction forming dense regions, whereas the Dark Matter (DM) lagged, forming a halo around our galaxies.

The alternate theory presented herein, is that only DM existed initially, and the concentrations of DM causes high concentration of energy (gravitational potential) which leads to the formation of our galaxy, and our sub-universe. This theory in one aspect is very similar to the theory presented by Olin Eggen, Donald Lynden-Bell, and Allan Sandage [7] in 1962, who proposed that disk galaxies form by gravitational collapse of a gas cloud, causing a rapidly rotating disk.

The Eggen theory has been abandoned, it explains formation of a large mass but does not explain the rotational disk shape.

The Eggen theory may be correct when (substituting DM for hydrogen) applied to a single DM concentration, which is basically near spherical. However, in order to obtain a rotating DM disk, two such DM concentrations are required to combine creating the high rotational angular momentum and resulting in a disk shape. The edge of this rotating DM disk is a very high-energy environment, causing the possible excitation of DM particles and creating our matter and our sub-universe. Thus our matter is created at each galaxy, and the SMBH flings our matter back into the larger universe. The DE is the original gravitational potential energy that is converted into kinetic energy during a galaxy formation, wherein the high-energy concentration acts to energize the DM particles.

2.1.4.1. **Our Matter Created at Each Galaxy**

DM is observed in the universe forming in large strings, and clumping around galaxies. Our matter appears to be created in high kinetic energy density regions, wherein colliding DMs excite their respective bubble structure, causing rise to an Electromagnetic force, and our matter.

Areas of high energy are thus the birthplaces for our matter, excited DMs. The observed concentration of our matter is in galaxies, which have a distinct disk shape. Colliding galaxies complicate the distinct disk shapes, and act to obscure the effects of the original disk shape. The main candidate for the region where our matter is created remains at the edge of a disk shaped Super Massive Black Holes (SMBH) in the center of disk galaxies. Our matter is possibly created in these very high-energy concentrations by exciting DM particle collisions at the edge of each galaxy's SMBH. DM material feeding a galaxy causes a continuous creation of our particles, and when a galaxy runs out of DM, it ceases to create excited matter and new stars. Each galaxy's excited DM is thus spread in a spinning disk shaped distribution determined by the SMBH spin. Thus the SMBH in the center of our galaxy is the mother of our galaxy. This theory explains the distribution of our matter compared to DM distribution in the universe, and explains why our matter is concentrated in galaxies. It also explains the more simple composite particles near the center of the universe. Fusion in the stars further build the composite particles into more complex atoms as currently described.

The formation of our matter in a galaxy is the universe's method to distribute the concentrated energy back into lower energy regions. Our excited DM is expected to decay back into the DM universe wherein the energy is converted back into gravitational potential between evenly distributed DM particles.

Other areas of high-energy concentration may also excite the DM material. Because a DM lacks the electromagnetic properties it can pass through neutron stars or black holes (BH) composed of our matter not being affected by the intense pressures inside a BH. In fact a DM will experience no gravitational force at the center of a BH or a neutron star because the summation of gravitational forces will null to zero. Similarly a DM without an EM force is not attracted or repelled by an excited particle, only experiencing its gravitational attraction. It is not contained inside of a star of BH, only influenced by the summation of gravitational attractions that can null.

However, these massive concentrations of energy may act to excite DM material, creating elements such as hydrogen inside the neutron star or BH, possibly causing it to explode in a super nova fashion. Thus even stars and BHs may be hydrogen fed by DM entering the stars and being converted to hydrogen. The energy concentration will make the BH unstable, and possible explode. Nature does not allow a high concentration of energy to be stable in a lower energy region. Thus stars may not be limited to their immediate hydrogen supply, but are re-supplied as the DM is converted into particles and hydrogen inside the star.

2.2. The Evolution of the Universe and Our Sub-Universe

The pattern of the universe is formed by two opposing phenomena. The first process requires DMs to collect together and transport gravitational potential energy into a small volume (sub-universe), which creates, excited DMs. The second process requires the high-energy regions to dissipate the energy back to low energy regions of space via the excited DM particles. The excited G_es decay back into ground state DM particles, which transfers the energy back into the ground state universe. Energy can only exist as associated with a DM that is as gravitational potential, kinetic energy, standing waves, or spins. Our matter may be the primary method excited DMs dissipate the energy back into low energy space.

2.3. Quiescent Infinite DM Universe

The steady quiescent DM universe is an even distribution of DM motionless particles where all energy is stored as gravitational potential. The gravitational fields from other DMs in the region independently affect each DM particle. In this quiescent state, any moving or excited DM, such as a photon, will disrupt the gravitational fields, which is a transfer of energy back into gravitational potential via gravitational field. Each disrupted DM will move under this gravitational field changes to reestablish a position wherein all kinetic energy is transformed into gravitational potential. Thus the kinetic energy is redistributed back into gravitational potential energy and the DM particles redistribute to their static motionless state.

2.4. We Exist in Both Universes

We live in two universes, one of normal matter and one of DM, sharing the same dimensions, time, energy and gravity. Only the DM universe is infinite, and our normal matter sub-universe has a beginning in time, an end in time, and a finite volume. Our sub-universe did not begin in a singularity or a big bang, but by a series of Super Massive Black Holes (SMBH) of DM. The SMBH at the center of our galaxy is our mother, not a potential destroyer.

2.5. Considerations for DM Feeding Stars & BH

2.5.1. Possibly Certain Size DMs are Excited

DM particles may also be of various sizes, being combined from the smallest fundamental DM particle. A specific energy concentration may be required to excite a particular size, thus the edge of a galaxy may only excite DM particles of a certain size, larger than Plank's constant. Perhaps in higher energy concentrations such as the center of a neutron star, smaller DM particles may be excited. A possible test of this theory may lie in observing matter near a neutron star versus at the edge of a galaxy.

2.5.2. BH May Only Exist at a Low Energy State

BHs may only exist at a low energy state; and at a low energy state, the DM particles may combine bubble structures until they become unstable and burst, transporting the matter away from a micro region. The very nature of a DM being a bubble structure prevents the collapse of matter. Plank's constant to the rescue.

3. A New Concept for Particle Physics

This bubble theory advances the concept that the excited DM particles provide the "particle" that when excited forms our matter. These excited DM particles create the electromagnetic force via their standing waves. The existence of the electromagnetic force differentiates the difference between DM and our matter (excited DM).

3.1. Our Matter and the Electromagnetic Force

The edge of this rotating DM disk galaxy is at extremely high-energy area, allowing for excitation of the single DM, G_g, into our matter, or into a G_e. The G_g is transformed into our matter possibly when the G_g bubble structure is excited, imparting an electromagnetic property. The excitation creates the Electric force field we observe which transforms the G_g into a particle like entity, a G_e. The electric force and the magnetic force are inseparable. This excited G_e forms our particles depending on what combination of energy excites the G_e.

3.2. New Particle Definition for DM and Excited DM

The entire universe is constructed of two particle families and energy. First a DM family, which earlier was defined as a gravity, bubble (G). Second, a family of excited gravity bubbles, which are our matter. G has a mass, thus all particles have mass, and the energy hosted on a G_e particle cannot be changed to mass.

The conservation rules for both G and energy independently are expected to hold. Gs cannot be converted into energy, but a G may host energy. Energy cannot be converted to a G. But energy can only exist associated with Gs. Gs have mass based on inertial and gravitational mass definitions. A massless particle is impossible because it would have to be solely composed of energy independent from being associated with a G particle. All energy must be associated with G particles. Energy cannot exist in any form without association with G particles.

All energy hosted on a G particle must be quantized except kinetic and gravitational potential energy. Thus matter is quantized, being G particles, and energy hosted on a G particle is quantized, but energy and distance are not quantized. Gravity is quantized.

3.2.1. The Fundamental Particle is a DM Particle

Current physic's theories consider the fundamental particles in two basic varieties: bosons, which have integer values of intrinsic angular momentum or "spin", and fermions, which have half-integer spin. Bosons include force-carrying particles such as the photon, W and Z and follow Bose-Einstein statistics. An important consequence of this is that many identical bosons are free to occupy the same quantum state, leading to phenomena such as Bose-Einstein condensates and lasing. Fermions are fundamental particles such as quarks and electrons, which obey Fermi-Dirac behavior. Identical fermions never exist in the same quantum state, creating a particle like shell structure behavior, which describes atoms, and, with it, chemistry.

This bubble theory defines the simplest particle to be a G particle of Plank's constant in size. The simplest particle of our matter is a G_e with the bubble structure excited. There may be a family of G_e when several G_e's combine. Perhaps the basic smallest G_e requires excessive energy to excite, not available at the galaxy's SMBH, thus not a common particle. Our basic particle may be larger than Plank's constant, based on available energy required to excite the bubble structure. However, other regions such as a concentration of our matter in super nova may create smaller G_e's.

The G bubble description avoids a singularity, and energy resonance also does not have singularities or points in space. This theory is similar to the Heisenburg uncertainty principle, by stating that matter cannot be described as existing at a point, but has a volume. Points are an artificial mathematical concept not shared in nature. Accordingly a G particle is not a singularity but exists over a volume. Similarly energy can exist only over a volume hosted by a G or composite G particle, or as a FF. All energy forms have volume and do not exist as points or singularities. A G's center is hypothetical, convenient for simplified math descriptions only. Singularities do not exist in nature in any form.

In contrast to current theory, all the G_es carry energy. Force particles do not exist. All particles contain mass. Particles are to be defined by their combination of energy excitation according to the allowable energy excitation degrees of freedom.

3.2.2. DM Particles Have Mass

The current definition of mass works very well, but the understanding of mass must change, since mass is composed of gravity bubbles with or without energy.

The current definition of mass must be extended to DM. Mass has properties of inertial mass and gravitational mass. Inertial mass is the kinetic energy required to provide a velocity to a particle, including a DM (G) particle. Since mass is quantized, the sum of Gs equals the mass in the Newtonian equation $K=1/2mv2$, $K=$ kinetic energy, $m=mass= \sum_n$ Gs. To be comfortable with this result one must accept that a G even in the DM state can host kinetic energy, thus has inertial mass, as well as gravitational mass. In the DM state mass acts "empty" since the DM particles can pass through each other. Thus mass is intrinsically non particle like, and empty except for its gravitational field.

The gravitational mass of a body can be determined by the summation of Gs in each body, which will also be the mass in current kilograms. Thus this DM particle in the field of gravity contains a mass having inertial and gravitational mass, but may be considered more of a force particle. However, it satisfies the requirement for mass, and must be considered to have mass.

The result is that a massless particle is impossible. All particles have mass, including a photon. The mass is not a function of velocity, and does not change with velocity. Mass cannot be converted into energy, and energy cannot be converted into mass. But mass can host energy, and energy cannot exist without mass. Both mass and energy are conserved in the universe and in our sub-universe.

3.2.3. Particles That Travel the Speed of Light

Einstein postulated that the lowest mass particle could travel at the highest velocity, and larger particles at slower velocities. Thus the photon was defined as massless and with the highest velocity. Thus mass was considered to be the limit for velocity. Einstein was unaware of DM.

The bubble theory presented herein considers a wave velocity to be a particle's velocity limit. This wave velocity limit requires a medium, which is identified as the gravitational field.

The DM particle exhibits kinetic energy, and by Force=(mass)X(acceleration) exhibits a inertial mass like behavior. The velocity of a DM particle may indeed only depend on available kinetic energy, and not be limited such as a photon with wave characteristics. A photon's speed, considered an energized DM particle, has a velocity defined as the pulsed wave velocity in the gravitational field. Thus a photon's velocity is limited to the wave velocity, and a photon is an energy wave mode on the excited DM bubble structure. A photon exceeding the speed of light is not possible as a wave in the gravitational field.

This gravitational field associated with every particle extends to infinity, filling the vacuum of space with a gravitational field. Space cannot be empty, always containing a gravitational field or a combination of gravitational and electromagnetic fields. The combination of a gravitational field and a electromagnetic field is referred to herein as a Force Field (FF). This FF fits the "ether" postulates.

3.2.4. Degrees-Of-Freedom to Excite DM Particles

There are multiple degrees of freedom wherein energy can be hosted on a G. These degrees of freedom are the alphabet soup for our matter and particle theory. The G_e's bubble like construction can support many combination of energy in the form of forward, left, right spin as well as three internal standing wave bubble resonance's (modes) on each potential shell. If two or more G_es can combine bubble structures forming a larger G_es, smaller G_es may be supported possibly trapped internal to the larger G_e bubble with similar properties, or exterior of each other via attractive forces such as gravity and the electromagnetic force. These G_e may also attract each other forming groups that create a composite G particle. Except for gravitational potential and kinetic energy, all energy is quantized when supported or associated with a G_e. Quantum Mechanics describes all G_e particle behavior. However, energy in itself is not quantized being in the form of kinetic or gravitational potential, which is another way of saying dimension, is not quantized. An analogy to the G_g is a ghost balloon, but of a non-particle like entity. An energized G_e acts particle like because of the electromagnetic phenomena created by the vibrating bubble structures.

The G_es may also vibrate in three dimensions, each mode having several DoF. A very large number of combinations become possible, each combination creating a particular particle. A paper by Y. C Chang and L. Demkowicz computationally describe the vibrations of a spherical shell, which is applicable to this theory for an excited G_e. [8] This theory allows very rigorous mathematical descriptions. Mathematics coupled with particle observations may possibly be the best method to prove or disprove this bubble theory.

Kinetic energy and gravitational potential are not listed as degrees of freedom because gravitational potential and kinetic energy are stored external to the particle. Both DM and our matter share these energies, but the ground state DM universe is expected to have a zero kinetic energy.

3.3. Naming Method For Excited Particles

A naming method should describe the particles similar to the method used in chemistry. A natural method is to identify the excited degrees of freedom by letters, and the allowed energy shells in a DM particle by the energy packet. Other required identifications are stable and unstable states, and composite particles. A simplified initial draft example is presented in Figure 2.

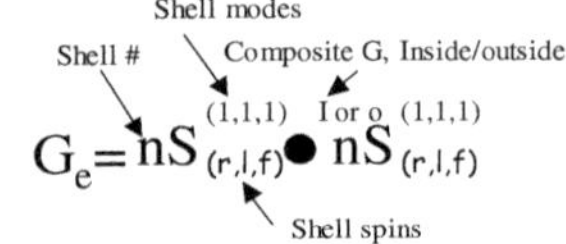

Figure 2 Proposed Particle Identification

3.4. DM Particle Discussion

These DM particles, G_gs are not particles in the current sense, but best described as bubbles, or 3-D strings. The Gs exhibit only positive gravitational attraction to each other, without a negative counterpart, and can host kinetic energy. As described earlier, a bubble is hypothesized in order to avoid the dilemma of a gravitational field approaching infinity as one approaches the center. Nature seems to abhor singularities. The minimum diameter of a fundamental G and an excited G_e is expected to be on the order of Plank's constant. Some limit must also exist to avoid a singularity.

The actual bubble structure may be more than a single membrane, but have some thickness with a variance in stiffness decreasing radially outward and possess infinite number of shells, with each shell supporting a quantized energy packet. The bubble structure describes a single shell. The actual particle is a nested infinite series of shells capable of supporting energy packets over a non-quantized range of values; however, each shell is a quantized energy packet.

Such a bubble structure description allows a photon to gradually lose energy to a gravitational field in a gradual non quantized manner. However, the allowed energy in a shell remains quantized, with the decreasing energy packet changing shells. Thus as a photon would "decay," its wavelength would become gradually longer, and the energy packet on the photon would shift radially outward to lower energy quantized shells. A description of the "tired light" theory was presented earlier in reference six.

Gs may also collect, held together by their respective gravitational force, both as a grouped collection of Gs and Gs whose bubble structure has combined into a larger G. These composite Gs provide a family of G sizes up to some unstable state, wherein the Gs burst back into smaller Gs. Relative kinetic energy between two intersecting Gs determines if the Gs combine or pass through each other by overcoming the gravitational attraction. This bubble theory also exhibits a weak and strong nuclear force nature. The DM bubbles can pass through each other since they do not exhibit a repellant force. The attraction between DM bubbles is also a function of position. Outside of each other's bubble they exhibit a stronger gravitational attraction similar to a strong nuclear force, but when passing through each other the attractive force is decreased because the bubble structure is the source of gravitational attraction, not the centers. Discussed in more detail later in the document.

Data supporting this hypothesis is observed in the Bullet Cluster, figure 3. The Bullet Cluster has two DM masses that passed though each other without velocity or position change relative to their respective galaxy, whereas a gas cloud of our matter interacted and stopped following its respective galaxy [9]. Based on that observation it is theorized that DM is a particle having a weak gravitational attraction that

Figure 3 The Bullet Cluster Composite [9]

allows mass like properties, such as hosting kinetic energy. However, the DM particles can pass though each other like ghost bubbles if their kinetic energy is greater than their gravitational attractions.

These G_gs in the ground state are very well described as ghost bubbles. The energy required to energize a very small G may be too great to occur in some environments, allowing for the possibility that a selection of Gs at certain size are energized to become G_es.

The implications for physics are that gravity is quantized. Mass is also quantized by virtue of being an excited or ground state G particle. The beginning of Quantum Mechanics originated in ancient times when mass was considered quantized. These G particles are what ancient Greek and Indian philosophers theorized as the fundamental particle, or atom as named by Greek philosophers. The term "atom" was prematurely assigned to our composite particles, the atom.

The G_gs' gravitational field extends to infinity in a manner wherein each G_g's gravitational field permeates the universe with a radially decreasing gravitational field.

The ghost like bubble structure for Gs in both ground and excited states act independently from the Gs' gravitational field in that the excited G_e bubble structure can spin, but the gravitational field does not spin as evidenced by our own earth's spin not dragging the gravitational field. A G is best described as a 3-D string exhibiting a gravitational attraction to a second G. All matter is composed of Gs in ground and excited states. Thus $E=mc^2$ is incorrect. Mass does not exist as convertible into energy, but can host energy, and mass can only be DM particles. The energy in an excited DM particle is the summation of the energy in each excitation mode, the spins, and the particle's kinetic energy. The G force particle is the fundamental particle for all mass, and can host energy internally in the excitation modes and externally in the form of kinetic energy and gravitational potential.

3.5. Physics of a Large DM Concentration

The physics of a large DM (G_g) concentration is counter intuitive, and very interesting. A large concentration of G_g will revolve around their common geometric gravitational center but in a different manner than we observe for our matter. The Gs can pass through each other, allowing all simultaneous orbits in all directions around this common center without effects of particle like collisions. The gravitational attraction on single Gs also varies based on the distribution of other G_gs. At the center of a symmetrical spherical concentration, the gravitational attractions null (zero), and the G_g orbit is a function of its kinetic energy and changing gravitational potential. Thus we have many oscillating G_g with all possible orbits first forming large strings, and then forming a spherical like configurations with the greatest G concentration near the outer edges. Thus creating the observed "DM halo effect" around galaxies. Two such DM concentrations upon merging will transform the gravitational potential into angular momentum, causing the new DM concentration to form in a disk shape with the exterior of this disk G_es approaching the speed of light, as observed. Velocities of our matter, excited G_e, have been measured at 84% of the speed of light; however, the G_g may very well have greater velocities and possibly exceeding the speed of light because unexcited DM particles are not subject to the wave velocity limit as photons. (Presuming the gravitational field is an FF limiting the wave velocity of on an excited G_e.)

3.5.1. No Pressure Inside a Large DM Concentration

A massive DM (G) spherical concentration does not form a container as our mass does. The mutual repelling force between our particles forms a shell that traps interior particles and creates pressure which traps and exerts a force on the interior material, transforming the summation of gravitational force of all of the material around a particle in the sphere toward the common center's gravitational center. The particles in a DM sphere can pass through each other, and are only affected by the summation of gravitational attractions from the surrounding DM distribution, which cancels out at opposite directions for each DM. Thus at the center of the DM sphere the gravitational force is zero. A DM sphere does not create pressure, or form a particle trap. At the center of our masses sphere, the gravitational force is the absolute value summation of all of the attractions from both sides, creating pressure. Whereas DM concentrations tend to have hollow centers in both the spherical and disk shaped concentrations.

4. Other Impacts On Physical Concepts

4.1. The Universe Requires a New Term for its Energy State

Energy laws similar to the thermodynamic laws are required. These energy laws must recognize energy in its many forms rather than just thermal energy, which is molecular kinetic energy.

The energy in the lowest state of the universe requires all particles to be at rest with all energy stored as gravitational potential energy. A particle with kinetic and stored energy will lose that energy to gravitational potential energy as it moves to the lower energy density regions in the universe. Temperature is not a good metric to describe the energy in the universe. Temperature for excited G_e is a measure of the particle velocities, rather than the energy stored in each G_e. Our matter forms a distribution of molecular velocities because of collisions, and subsequent trading of kinetic energy. DM does not share its energy via collisions, only via the gravitational field. Thus ground state in the DM universe requires motionless DM particles wherein all kinetic energy has been changed to a steady state gravitational potential. Gravity exerted on each DM is zero because of the evenly distributed DM particles. However, energy exists stored by the gravitational potential between each DM particle. The energy is evenly distributed at the universe's lowest energy state. A disturbance to this steady state gravitational potential can be visualized by moving a single DM, which causes a 3-D gravitational potential change, which influences adjacent DMs causing them also to move and emanate 3-D gravitational field changes. Each DM seeks the lowest gravitational field state, thus the DMs collectively distribute themselves evenly throughout the region. The gravitational potential changes act to transfer the kinetic energy equally and redistribute the DM particles transferring all of the energy into gravitational potential. These gravitational field changes are not the classical waves because a wave carries energy directly. These gravitational field changes are analog like functions that redistribute gravitational potential energy by changing the gravitational field potentials. Each DM particle seeks the lowest gravity field potential, which is an even DM distribution. This lowest energy state is contrasted to our familiar universe with excited matter wherein unbalanced gravitational and EM forces exist, as well as high kinetic energy particles and energy stored as excited G_es. Intergalactic space in our universe contains approximately one hydrogen atom per cubic meter, as well as gravitational and electromagnetic fields. The contrasting DM deep universe is an even distribution of G particles at rest with no measurable gravitational field at each G, and no electromagnetic forces.

4.2. Electromagnetic Force and Weak-Strong Nuclear Force

The electric charge, and resultant electric force is theorized as a product of energized G_e by quantum energy packets.

The magnetic force and the electric force are inseparable, and associated with each other. Which excitation mode creates the EM force requires further study. However, the end result is that the energized G_e exhibits particle like behavior.

There is a possible explanation for the weak and strong nuclear force. A standing wave on the G's bubble structure is theorized to create an electric charge. This electric charge can be mathematically treated as emanating from the center of each G at a distance, but should be treated as emanating from the bubble surface at close distances. The attractions between the two Gs decreases as the Gs enter each other, causing a drop in the gravitational and electromagnetic attraction, or a weak nuclear force. Thus a strong attractive nuclear force weakens as the two particles enter into each other. The arrows in figure 4 demonstrate the concept by showing a sample of the canceling forces. If a second G is smaller than the first, it may enter the larger G, and if its quantized kinetic energy is allowed, remain inside the first G. The G's bubble structures may also combine, creating a pocket wherein repellant forces between particles, such as protons, are nulled, a possible model for an atoms nucleus. An energized larger G may indeed capture a smaller G, or combine forming a larger bubble structure capable of hosting longer excitation modes. Thus a possible model for the matter we are familiar with, and a possible structure of our atoms.

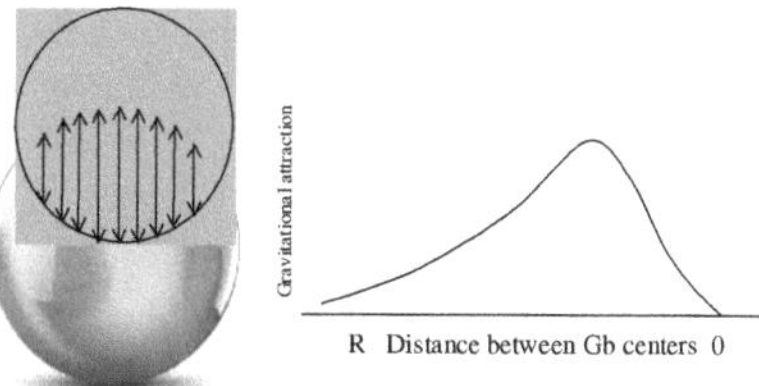

Figure 4 Reduced Attractions as Bubble Structures Merge

4.3. **Photons Being Both a Particle and a Wave Explained**

No immediate contradiction between a photon being both a particle and a wave. A photon is not hard particle but energy hosted on a G_e. The photon possesses the mass of its' original G_g particle. These are the simple conclusions; however, much is still to be understood.

Why do photons have a velocity the speed of light and other particles do not have that velocity? Why cannot photons be a particle when at rest? This theory predicts that all our matter consists of excited G_es, which also suggests that all particles should travel at the speed of light. Why is any particle at rest?

If the gravitational field is theorized to behave as an ether, and photons are "waves" in that field. Photon velocities will behave similar to sound in a medium, but unlike sound waves in air, the waves do not propagate in three dimensions outward. The waves are restricted to boundary conditions, the particle's energy quantum level for the particular shell in the bubble structure. Thus the wave exists inside or on the boundary, allowing for a pulse wave velocity, to propagate in one dimension in a limited region constrained by its boundary, and transporting the photon at the permitted wave velocity. This describes a photon. The next question is why can a particle be at rest? Can a G particle that is not excited travel faster than the speed of light?

4.3.1. Speed Of Light In a Vacuum Dependent on Force Fields

The gravity force that permeates the entire universe has been postulated as the possible "ether." Historically the ether is disputed, but not the gravitational field. The argument may be more of a semantics issue. Such a theory may explain how photons travel, and the speed of a photon. The observed speed of waves in water may be more than an analogy if gravity is considered the "ether." The speed of light is considered to be constant similar to sound in water; however, the speed is expected to be dependent on the gravitational field strength as the speed of sound is dependent on density. Such a theory treats light as a wave, and a disturbance in the gravitational field as carried by an excited G_e. An earlier flaw with ether theories is the presumption that the speed of light was a function of the source velocity similar to a baseball thrown from a moving vehicle. Photons are waves, and will behave as a wave in the gravitational ether. The wavelength changes, not the velocities, for both sound and light when generated from a moving source. The velocity of light is independent of the speed of its source, but dependent on the absolute value of the gravitational field strength and possibly the electromagnetic field as evidenced by the Kerr effect.

Another consideration wherein a photon's velocity differs from a wave in water is based on the physics of the waves. In water, the wave is an exchange of kinetic energy between water molecules, causing a traveling kinetic energy wave. A photon travels in a massless energy FF with its wave localized inside of the photon's bubble structure. A photon in matter can travel similar to the wave in water by being absorbed and reemitted by atoms. These are similarities, but the physics is different.

R. Battesti, C. Rizzo published a very complete history of experimentation investigating the speed of light varying in a vacuum when subjected to EM fields. The experiments have demonstrated that a vacuum contains an induced index of refraction when EM fields are applied. [10] The paper attributes the phenomena to Dirac's virtual electrons. However, another explanation is that the EM fields act as a force field similar to a gravity field. These experimental results may be the smoking gun proving that fields do act as the ether.

The theory presented herein also predicts that only wave energy is limited to the speed of light. Mass may travel faster than the speed of light, but waves are limited by their wave velocity in the gravitational field or FF field. The DM particle contains mass and an un-energized DM particle's velocity may be only limited by its kinetic energy. Whereas an energized DM particle's velocity is limited by its wave velocity in the gravitational and/or EM field. The G particles are not restricted by the wave velocity restrictions, as G_e particles are restricted. It appears no limit exists except for required kinetic energy, which becomes unlikely at some point. A high kinetic energy G particle is also highly at risk of becoming energized, which would limit its velocity to the wave pulse in the gravitational ether, or the speed of light. Perhaps a G faster than the speed of light has a short lifetime velocity, being converted to an energy carrying G and accordingly the wave function limits the velocity to that of light. Note that this velocity limit changes as the gravitational field strength varies.

4.4. Gravity is Quantized

Gravity is quantized, being a single G_g. The gravitational field experienced by a single G is a summation of all of the contributing G_gs and G_es, which add and subtract dependent on the proximity and arrangement of other G_gs. The gravitational field is the absolute values of the same sum.

The gravitational field from each G exists to infinity in a rapidly decreasing value based on the newton (N) in classical Newtonian physics. The effects of changes in position of each G must propagate through this field, and are most likely analog changes in the gravitational potential values rather than waves. A change in one DM particle's position causes every DM particle in the area to also react changing its influence on all of its neighbors. The gravitational force transfers kinetic energy away from each G, into a continuum between Gs as gravitational potential, a perfect description of DE. Gravity waves it appears cannot exist. However, an alternate explanation for recent observations cannot be provided. Force change is propagated on the changing gravitational field, not energy. The velocity of this force change is not limited by a wave velocity, and may be instantaneous.

Relativity seemed to answer the observations of planet position and orbits, which were not perfectly described by Newtonian physics. However, DM was unknown at that time, and would have a substantive effect on orbit predictions. In fact DM is theorized based on orbits not following expected trajectories, and gravities refraction of light. Any correction provides a better fit to observations, even though the reason for the correction may be bogus.

4.5. Energy and Mass Conversion Disputed

This theory disputes the mass conversion to energy in General Relativity by maintaining that mass and energy are not interchangeable. Mass exists as either an excited or ground state DM particle, and cannot be changed to energy. The conservation of energy and conservation of DM gravity particles hold. DM particles can host energy, by transforming into an excited DM particle and becoming our familiar particles. But mass cannot be changed to energy and energy cannot be changed to mass.

Einstein's $E=Mc^2$ is a casualty. This concept should be replaced with the concept that a G exhibits gravitational force and mass as well as inertial mass. But energy is independent and can be stored between Gs as gravitational potential, kinetic energy or on an excited G_e. The energy in a G_e particles can be converted between gravitational potential, kinetic energy, or excited G_es by several DoF such as spin or modes on the bubble structure. Fission is an example of breaking up composite particles and freeing some of this energy into kinetic energy and other particles such as photons. Fusion has a similar effect, except not all of the composite particles are used to make the larger composite, the leftover changed back into kinetic energy and photons. Thus mass is defined as a DM particle which cannot be changed to energy, and energy cannot be changed into a DM particle. The host G_g cannot be converted into energy. The conservation of G_g and energy is expected to hold. The Einstein equation explains the massive energy available in a G_e; however, an excited G_e provides an alternate theory for existence of energy on the particle. All matter is composed of two items, Gs and energy, and the composite creates our matter.

4.6. Cosmological Constant

The Cosmological Constant (CC) is the stored gravitational potential between gravity bubbles in a DM universe region of space. The lowest energy state for the CC is the DM universe wherein all Gs are at rest, with gravitational attraction being cancelled by the symmetry of the G distribution. The energy is directly related to the G distribution density.

In DM regions of our universe, where the G's (gravity bubbles) concentration is greater, the CC is higher. However, in regions of our universe (an area of excited G_es) the energy stored on each of the G_es must also be considered.

4.7. No Big Bang

Nature abhors singularities. The conversation of Gs and energy holds, which means neither, can be created from nothing. The DM universe exists as a soup of DM entities evenly distributed and motionless, extending to infinity in three directions, and in past and future time to infinity.

Fluctuations in this DM pea soup cause concentrations of DM, forming massive DM bodies that act to pool potential gravitational energy. This potential energy can be converted to different forms of energy, such as kinetic on a DM particle, and excitation of a DM particle.

Our universe is created by DM fluctuations which first form galactic strings of DM. Inside these DM strings, further spherical like concentrations form. These concentrations pool energy and form massive spherical halo type masses. When two such DM masses converge, the gravitational potential energy is converted into disk shaped DM concentration with high rotational energy. The edge of this disk can approach the observed 84% the speed of light for atoms of our matter. This disk energizes the G_g into our G_e entities, which exhibit a particle like behavior. The birth of our matter is thus the SMBH at the center of our galaxy. Our matter is continually generated by the galaxies SMBH and flung into our disk shaped galaxy. This matter transports the high concentration of energy back into the universe.

Galaxy clusters thus form inside the galactic sized strings. The form of our matter may be determined by the energy

source. Plank's constant may describe our current smallest
particle based on the energy level required to excite a G. Thus
Plank's constant may be the minimum size of an excited DM
particle, with particles created at SMBH being larger.

The expectation is that the G_g may have velocities greater
than the speed of light because it is not a wave, and not
limited by wave velocities, only available kinetic energy.
However high velocity G_g upon colliding may energize the bubble
structure creating our matter whose velocity will be limited by
the wave velocity, possibly creating a photon.

4.8. A Star's Gravity Effect on Passing Photons

Einstein's General Theory of Relativity states that matter
and energy actually mold the shape of space and time dilation.
The "force" of gravity is simply the sensation of following the
shortest path through curved, four-dimensional space-time. Space
and time are variables and interlocked to each other, while
matter and energy are interchangeable. Thus explaining the
gravitational lens effect on a photon passing a massive body.

This Bubble Theory offers a different expiation. Red shift
is observed from light leaving our sun, indicating a
gravitational effect that subtracts energy from the photon,
while maintaining the photon's velocity. Literature sources
refer to gravity as creating an index of refraction to explain
light bending in a gravitational field. This index is more of an
empirical conclusion rather than a derived result. Providing
gravity with an index of refraction can obtain the correct
answer by merely defining the index of refraction at the correct
value. A derived equation is preferable, which is possible using
Newtonian laws for the gravitational field Newtonian mass
influence on the passing photon as well as the gravitational
field's non linear optical effects on the passing photon. The
photon's path should be calculated by considering the
gravitational field acting as both a Newtonian result of mass
gravitational attraction and an optical FF with an index of
refraction between the photon's mass and the star's mass.

The bending of a photon passing a star can be accurately
modeled without concepts of space bending or an empirical index
of refraction. The photon's path is a combination of three
contributors, Newtonian physics on the photon's mass, wave
velocity index of refraction caused by the Force Field (FF)
which is a combination of gravitational and EM fields, and an

index of refraction from the gaseous cloud extending around the star.

A derived result calculates according to Newton's law the gravitational attraction between the composite G_e photon particle's mass and the star's mass. The energy carried on the G_e photon is not part of the attractive force, and cannot be handled as mass. Thus a G_g and a G_e will be attracted exactly equally to a star based on their mass, and described by Newtonian physics plus from the gravitational field's index of refraction. A photon appears to be a composite of G_es and energy with the G mass component behaving according to Newtonian physics when passing a massive body.

5. Further Exploration Required

5.1. Why Do Only Photons Travel at The Speed of Light

Why do only photons travel the speed of light and other particles can be at rest? A possible explanation is that all particles travel at the speed of light since they are all composed of waves in the FF; however, a photon's G_e excitation's result in one direction displacement in a single dimension, whereas other more complex composite particles excitation's are in all three dimensions in all three directions causing the composite particle to vibrate in place at the speed of light, allowing for seemingly at rest particles and lower velocities. The displacement nulling at the particle level is similar to Brownian movement; except, Brownian movement is molecular collisions and exchange of kinetic energy whereas opposing wave velocities in the gravitational FF causes the composite particle to dither in place.

Thus a photon's excitation results a single dimension displacement with the pulse wave velocity propagating at the speed of light, in the same direction. The photon is expected to host other energies such as kinetic energy and spins.

Another possibility is that the excited G_e bubble structure is only excited in two modes, $(1,1,0)$ allowing for a cross product creating a result similar to the Poynting vector of Electric field cross magnetic field. When all three modes are excited, $(1,1,1)$, the particle appears stationary, but oscillates in place.

The photon and its energy can only an absorbed wherein the absorbing particle is capable of nulling the photons velocity and apparent existence as a single photon. Einstein postulated, "the lightest particle has the greatest velocity, and a photon was considered massless. Einstein's relativity theory states particles with mass gain mass as they approach the speed of light, limiting their velocity to sub light velocities." Neither of Einstein's postulates is accepted herein. A photon has mass and energy, and the photon's mass consists of its G particles, and the photon's energy is its kinetic and excited mode energy plus possible spins. A composite particle maintains the same G particles at rest or at any velocity, and does not change mass. Energy and mass are not interchangeable, but fixed by conservation of mass and energy laws.

William H. Cantrell, Ph.D. writes in A Dissident View of Relativity Theory, paraphrased "The photon's velocity will change as the photon enters our own solar system, and continue changing as it enters Earth's gravity. [11] The photon's velocity change will be gradual as it transits through the different gravitational fields. Such a theory is intriguing and "makes-sense." The theory also correctly explains data results from the Pioneer 10 and 11 deep-space radio data (and probably with the Venus radar data from the 1960s). (Also see *Infinite Energy* #52, pp. 33, 36.)" [11]

5.2. Photons May Decay

The decay of a photon (a G_e particle) back into G_g can be explained as the transition of a photon's energy into the gravitational potential. Thus a photon's wavelength shift would continue shifting until the photon's kinetic G_e energy was fully transferred back into the gravitational potential of the DM universe, and the photon eventually decays and reverts to a G_g because it cannot exist without its excitation mode, which provides its characteristic velocity. This process is expected to be the major method our energized sub-universe reverts back into the infinite DM universe, consistent with the law of thermodynamics wherein energy flows to lower energy states. This concept was referenced earlier as "tired light" in reference six. Such a theory also suggests an "ether."

Robert B. Laughlin, Nobel Laureate in Physics, endowed chair in physics, Stanford University, commented the following, paraphrased.

Einstein's general theory of relativity conceptualizes space as a medium; however, his first assumption in special relativity was that no ether existed. The word "ether" is ridiculed in theoretical physics because it opposed relativity, but very nicely describes the vacuum. Relativity does not predict the existence or nonexistence of matter in the universe, only requiring the matter to have relativistic symmetry. It turns out that such matter exists. Later studies of radioactivity showed empty vacuum of space exhibited spectroscopic structure, or an ether similar to ordinary quantum solids and fluids. Subsequent studies with large particle accelerators have now led us to understand that space is more like a piece of window glass than ideal Newtonian emptiness. It is filled with virtual particles that are normally undetectable but are detected in particle collisions. Experiments confirm the vacuum of space is relativistic ether. [12]

5.3.   Propagation Velocity for Gravity & EM Forces

The gravitational field associated with each DM exists in space, and the DM particle has always existed; therefore, there is no initial velocity of propagation issue. However, movement of these particles induces a change in the gravitational field values, and these changes may have a propagation velocity. A question is what is the velocity of these field changes as propagated through space? Similarly a newly created electromagnetic force created by energizing a G_g must have a propagation velocity. These velocities are not waves, and may not behave in the same manner as a photon. A change in the field strength may not propagate as a classical wave. The change may be similar to water rising or falling in a bucket, and could conceivably exceed the speed of light. The velocity of propagation of the field strength change is an open question.

5.4. Vacuum Energy

A current concept of the Vacuum Energy (VE) is that it consists of a sea of virtual particles or vacuum fluctuations. These particles appear out of nothing with opposite polarity, and annihilate each other back into nothing, thus conserving matter on the larger average. The excited DM particle theory is an alternate explanation for the Dirac's sea of virtual particles.

The Bubble Theory offers a different explanation in that the DM particles are detectable only from the weak gravitational field, therefore for all purposes undetectable. If a DM particle is energized it suddenly becomes detectable. The energy on the excited DM is quantized. Paul Dirac, based on mathematics, conceived empty space to be filled with virtual particles forming out of nothing and disappearing back into nothing, called the Dirac Sea. Dirac's virtual particles were electrons with negative energy. Dirac's theory is derived from mathematical formulations. [13] Dirac's concept is a very unsettling because it deviates from conservation laws in the short term. The "Bubble Theory" follows the conservation of mass and energy on the micro as well as on the macro average.

A DM particle is currently impossible to detect because of its very weak gravitational field. If this DM particle is energized, and provided the electromagnetic force, it would suddenly be detectable, and appear as a particle that appeared from nothing, and if the particle were unstable, decay almost immediately. However, based on a conversation of both energy and G (DM) particles, no conservation law is violated at any time, whereas the Dirac Sea presumes particle and anti particle are created which annihilate, and therefore conserve matter over the larger average. This theory provides a possible physical example and explanation for Dirac's sea and VE.

Current observations of a vacuum may indeed support the Bubble Theory, in that the vacuum has been identified spectroscopically as behaving like a piece of glass rather than an empty volume (Robert B. Laughlin). Since one tends to find what one looks for, future particle research at CERN may provide corroboration of a DM particle being converted into our matter. An understanding of particles that are stable and unstable may be required, expecting a very high number of unstable particles as compared to far fewer stable particles. A G bubble structure resonance may support higher modes, but only the $(1,1,1)$ mode would be stable, and several bubble shells may exist. Bombarding a particle may create any combination of breaking a composite particle into it's constitutes, or create even other particles. The energy level may also determine what G particles can be excited. The energy levels at the edge of SMBH or inside bodies of our matter such as quasars or neutron stars may be required to create our basic stable particles.

5.5. De Broglie Waves

De Broglie waves are not explained. All matter is composed of energy hosted on Gs. Photons have a wave and velocity in the FF medium. However why do slower moving particles have shorter waves and travel slower than the wave velocity? The answer may only be found by mathematically modeling the phenomena. Since all matter is energy on a G_e, quantization and wave like motion is expected, and Gs interacts with the gravity flux (ether) permeating the entire universe. Possibly near a composite particle the FF is high enough causing the wave velocity to be slower.

6. Conclusion

The presumption that a DM particle can become excited and create our particle dramatically changes our concepts in physics. However, it also creates a "make sense" theory that lends itself to a much more rigorous mathematical modeling than the current presumptions of a big bang and relativity. The contradictions of a photon being both a particle and a wave are also resolved. Nature does not have singularities, which means a particle is more accurately modeled as a 3-D string, or more specifically, a gravity bubble with infinite number of shells with an inner boundary. Such a concept is very well described by Quantum Mechanics. Particle Physics, Cosmology and Quantum Mechanics together may be just be redefined as General Physics.

This theory simplifies physics. The universe is created from only two items, rudimentary DM particles and energy. The possible combinations are much greater than a binary number because DM can combine and group forming composite DM particles. Energy can take many forms on these composite DM particles.

---

7. References:

1. Credit: ESA/Hubble, NASA, HST Frontier Fields.
Acknowledgement: Mathilde Jauzac (Durham University, UK) and
Jean-Paul Kneib (École Polytechnique Fédérale de Lausanne,
Switzerland).

2. NASA/JPL PASADENA, Calif. – Two X-ray space observatories,
NASA's Nuclear Spectroscopic Telescope Array (NuSTAR) and the
European Space Agency's XMM-Newton,. "NASA's NuSTAR Helps Solve
Riddle of Black Hole Spin" NuSTAR, Feb. 27, 2013

3. Andy Goulding and Prof. Jenny Greene of Princeton University
"Connecting Dark Matter Halos & Supermassive Black Holes" at
website < http://www.astro.princeton.edu/~goulding/>

4. Akos Bogdan, Harvard-Smithsonian Center for Astrophysics as
reported by Feb 20, 2015 by Editors of SCI-News.com

5. John Wheeler, 1955 From Wikipedia, the free encyclopedia

6. Pratt, David PhD "Cosmology and the Big Bang,"
http://www.theosophy-nw.org/theosnw/science/prat-bng.htm, pp.
10-11, citing Van Flandern, Thomas PhD. Dark Matter, Missing
Planets and New Comets, North Atlantic Books, 1993 and
LaViolette, Paul PhD. Genesis of the Cosmos, Bear & Co., 2004.
See also Van Flandern"s Article "The Top Thirty Problems with
the Big Bang" on his website www.metaresearch.org/cosmology.

7. Allan Sandage. From Wikipedia, the free encyclopedia.

8. 632Y. C. Chang and L. Demkowicz, "VIBRATIONS OF A SPHERICAL
SHELL. COMPARISON OF 3-D ELASTICITY AND KIRCHHOFF SHELL THEORY
RESULTS." The Texas Institute of Computational and Applied
Mathematics, The University of Texas at Austin, Austin, Texas
78712, May

9. 1994The Bullet Cluster Composite Credit: X-ray: NASA/CXC/CfA/
M.Markevitch et al.; Lensing Map: NASA/STScI; ESO WFI;
Magellan/U.Arizona/ D.Clowe et al. Optical: NASA/STScI;
Magellan/U.Arizona/D.Clowe et al.

10. R. Battesti, C. Rizzo, "Magnetic And Electric Properties Of
Quantum Vacuum" Physics Optics, 5 Nov., 2012

---

11. William H. Cantrell, Ph.D. writes in "A Dissident View of Relativity Theory," *Infinite Energies* #52, pp. 33, 36.

12. Robert B. Laughlin, Nobel Laureate in Physics, endowed chair in physics, Stanford University. From Wikipedia, the free encyclopedia.

13. Paul Dirac, From Wikipedia, the free encyclopedia.

About the Author

Daniel Stark was born in Peach Valley, near Delta Colorado, on November 18, 1943 in a one-bedroom ranch house. In 1952 the family relocated to Acampo, California forced by bad economic times.

Daniel put himself through college working summers. In 1968 Daniel graduated from Fresno State College with a BS in physics,(Dr. Scott's words to the physics graduating class, "remember, you have the same physics training as Einstein.") was drafted for Viet Nam, and married a Barbara Boswell. Daniel was assigned to Navel Weapon Systems Engineering Station (NSWSES) as a civilian physicist. NSWSES provided engineering aid to the Navy and Test & Evaluation support for developing systems. NSWSES offered an action-oriented environment with all programs being a DX1 (in battle) priority, and a can-do attitude. Daniel provided Electronic Warfare expertise for the DD class ships. A later assignment was test conductor for Close in Weapon System (CIWS), also called Phalanx. Dan developed a cylindrical corner reflector replacing the nose cone on a 5"/54 AAC shell which with its enhanced radar cross section, became a test target for CIWS testing. The enhanced radar reflection shell fired from an offshore ship at the CIWS system simulated a missile, providing an inexpensive target.

In 1979 Daniel joined GTE Sylvania optics group in Sunnyvale, CA. as a section leader for production of laser radar called PATS.

In 1983 Daniel joined Lockheed . Later he transferred to the Palo Alto Advanced Test Center (ATC). ATC offered an environment with a variety of developing programs and brilliant personnel. He specialized in "red programs" wherein the system was ready for delivery, but did not operate, and behind schedule. These situations were an exercise in logic wherein the given truths were typically incorrect, and one needed to technically understand the system and identify who could resolve the design issues. It was a stressful but highly satisfying experience when completed.

Upon retiring Daniel has pursued inventing, and currently has four patents with several additional planned.